George Sumner Huntington

The Significance of Muscular Variations

George Sumner Huntington

The Significance of Muscular Variations

ISBN/EAN: 9783337715151

Printed in Europe, USA, Canada, Australia, Japan

Cover: Foto ©berggeist007 / pixelio.de

More available books at **www.hansebooks.com**

The Significance of Muscular Variations, Illustrated by Reversions of the Anti-Brachial Flexor Group.

By GEO. S. HUNTINGTON.

[Trans. N. Y. Acad. Sci., Vol. XIV. (Aug. 28, 1895) 231-259.]

THE SIGNIFICANCE OF MUSCULAR VARIATIONS, ILLUSTRATED BY REVERSIONS OF THE ANTI-BRACHIAL FLEXOR GROUP.

By Geo. S. Huntington.

Read by Title, Feb. 11, 1895.

The study of muscular variations, if carried on systematically, embracing observations made on a large number of subjects, cannot fail to reveal certain finer differentions, which, while lost in a mere enumerative record of muscle variations, gain a new and important significance when grouped together and compared in an effort to trace the morphological meaning of the variant condition. This becomes most strikingly apparent in the case of certain appendicular muscles, and especially of some muscles of the fore-limb. The extreme modifications which, in vertebrates possessing an anterior extremity, widely different functional requirements have impressed on this portion of the locomotory apparatus, may properly be held responsible for the fact that here variations of the most composite type are to be encountered. A glance at the divergent forms presented by the vertebrate pectoral girdle will convince us that the muscular structures connected with the same must offer deviations from the primitive types, which not only change the arrangement of homologous muscles in different forms, but which will afford the opportunity for numerous reversions to the original condition in the case of any individual muscle.

The specialization of the fore-limb, in exchanging a purely or chiefly locomotory function for one which includes or substitutes the prehensile function, and the consequent higher development of the manus, has more especially served to single out the anti-brachial flexor group, and to add to the function of flexing the forearm at the elbow-joint the more complicated movements of radial rotation in supination of the hand. In some instances this additional functional requirement has sufficed to specialize certain members of the flexor group as supinators, modifying the insertion so as to restrict the same chiefly or entirely to the radius, and separating a portion of the flexor mass by complete cleavage from adjoining muscular strata.

This is most excellently seen in the so-called Biceps flexor cubiti of Man and Primates generally, a muscle which acts as one of the chief radial supinators, while its flexor function is to a considerable extent subordinated to this main action.

In muscles of this type variations are frequently encountered, many of them being reversional in character and pointing to previous ancestral types of myotecture.

The definition, which certain muscular attachments obtain by specialization of function, is accompanied by the concomitant reduction or elimination of other connections which have lost their original value and significance in the new sphere of the muscle, or would even, if retained, interfere with the same.

The muscle, in returning in an incomplete form to these obsolete conditions, presents reversions to which many of the observed variations may properly be ascribed.

No muscle presents these features more strongly developed than the Biceps flexor cubiti. The muscle is, next to the Palmaris longus, the most variable in the body, Macalister* in his Catalogue of Muscular Anomalies enumerating no less than 45 separate variations.

It is our purpose in the present paper to discuss only a certain group of these variations, and to endeavor to obtain an insight into their significance.

The human material, on which the following observations have been made, consists of 464 upper extremities dissected during the year 1893 to 1894 in the Anatomical Laboratory of Columbia College.

As the explanation of the probable derivation and significance of certain of these variations of the Biceps muscle is closely connected with the morphology of the remaining antibrachial flexor group, it will be desirable to take a general view of the arrangement of this group, before proceeding to the details.

In the lowest vertebrates a continuous and non-segmented plane of muscular fibres proceeds from the ventral aspect of the trunk to the flexor surface of the anterior limb. This condition is best represented by the muscular apparatus of the pectoral fin in fishes.

In ascending the scale differentiation leads to the more or less complete separation of this muscular mass into layers and strata, which may or may not preserve connections with each other, indicative of their original union.

A consideration of the general arrangement of the flexors of the forearm in higher vertebrates reveals the existence of three main divisions or layers,† more or less blended with each other.

---

*Alexander Macalister, Additional Observations on Muscular Anomalies in Human Anatomy (III. Series), with a Catalogue of the Principal Muscular Variations hitherto published. Transact. of the Royal Irish Academy, Vol. XXV., Pt. I., Dublin, 1872.

† G. M. Humphry, Observations in Myology, Cambridge and London, 1872, p. 163.

1. *Superficial layer*, a portion of the superficial ventro-appendicular muscular sheet, reaching, usually by combination with the succeeding layer, a skeletal or fascial insertion in the forearm. Usually but few and unimportant traces of this layer are found in higher forms, the great bulk of the superficial fibres passing from the trunk to the anterior extremity terminating, as Pectoralis Major and Latissimus dorsi, respectively on the radial and ulnar margin of the humerus. Occasionally, however, the Pectoralis Major is not arrested at the radial tubercle of the humerus, but, as in Orycteropus, accompanies the Biceps to the radius, or, as in the Otter and Wildcat, together with some fibres of the Trapezio-Deltoid, accompanies the Brachialis anticus to the ulna, or, as in the Seal, expands into the fascia of the forearm and thus reaches the hand.*

2. *Intermediate layer*, constituting the "extrinsic" limb muscles of Humphry, derived from the deeper portion of the ventro-appendicular sheet, forming muscles arising from the pectoral girdle and passing over the elbow to the bones of the forearm. This group is represented by the Biceps, which muscle, besides its relation to the Coraco-humerals, presently to be considered, presents various degrees of continuity with, and separation from, the following layer.

3. *Deep layer*, formed by the "intrinsic" limb muscles, arising from the humerus, and pursuing the same course as the 2nd layer, represented by the Brachialis anticus, or in some instances (Hippopatamus) by the Brachio-radialis†, which muscle in this animal is larger and occupies the space on the outer side of the humerus from which the Brachialis anticus usually arises.

Here we meet with a replacement of the usual Brachio-ulnar by a Brachio-radial flexor, to the complete exclusion of the former.

The 2d and 3d layers, *i. e.*, Biceps and Brachialis anticus, either considered by themselves, or more especially when viewed in the light of their mutual relation, present many points of morphological interest, and suggest explanations of human revisional variations of the former muscle.

A general review of the conditions found in the anti-brachial flexor group will call for the consideration of the following points:

I. *Morphology of the Biceps, the variations of the muscle and their significance.*

---

* Humphry, op. cit., p. 147.
† Humphry, op. cit., p. 163.

II. *Relations of Biceps to the adjoining deep ventro-appendic-ular muscles, viz.: the Coraco-humerals.*

III *Relations of Biceps to the deep intrinsic flexor of the forearm, viz.: the Brachialus anticus.*

I. MORPHOLOGY OF THE BICEPS FLEXOR CUBITI.

We may in the first place consider the composition of the human Biceps in the light derived from the comparative study of the muscle in other vertebrates.

Although our knowledge of comparative myology is still somewhat meagre, considering the extent of the subject, and although the results of investigations appear at times too fragmentary and uncertain, yet enough is known to justify us in regarding the form in which the Biceps usually presents itself in man as the result of certain specializations of function acquired by the upper extremity, which have brought about a greater definition of certain portions, whereas other parts, lacking the stimulus of this functional requirement, have retrograded and have become to a great extent rudimentary or entirely eliminated.

A. Origin.

In the complete form the Biceps occupies two points of origin from the shoulder-girdle corresponding to the anterior edge of the glenoid socket and to the coracoid element.

The actual arrangement of the muscle in individual forms will be naturally greatly influenced by the structure of the girdle, and notably by the predominance or reduction of the coracoid and its appendages. An instance of this is afforded by the extensive Coraco-radial muscle which represents the Biceps in Cryptobranchus japonicus, whereas, in many other forms, the coracoid origin of the muscle is much reduced or absent, and the Biceps appears as arising solely from the glenoid margin.

A glance at the arrangement in the vertebrate series will immediately indicate that the origin of the muscle is confined, strictly speaking, to the coracoid element of the pectoral girdle. It may arise by a single head, or, as in man, the origin may be double, hence the name.

If, as in us, the coracoid is reduced and the scapula correspondingly increased, the muscle, if it preserves its double origin, will arise by one head from the small coracoid process, by the other from that portion of the glenoid margin which is contributed by the coracoid bone. If the long tendon of the glenoid head of our Biceps is examined in subjects between the ages of ten and puberty, it will be found in every instance connected with the so-called sub-coracoid centre of ossification.

Hence our division of the muscle into glenoid or scapular, and coracoid heads, while convenient as corresponding to adult conditions, should not cause the fact to be forgotten that the Biceps is entirely a coracoid muscle in in its origin. We might, considering the probable significance of our coracoid ossification, speak of the coracoid or long and precoracoid or short head of the muscle. The origin of the Biceps taking place in this manner from a single skeletal element, it is not surprising to find the consolidation of the two heads, which are present in man and the primates generally, to be of very frequent occurrence in lower forms. It will then depend upon the prominence of the coracoid element, and the relation of the origin to the glenoid socket, whether this single head will -be denominated as the coracoid or glenoid muscle.

B. Insertion.

In like manner we find a considerable range in variation in the insertion of the muscle. It may be attached to either or both bones of the forearm, according to the functional character of the limb, and the specialization of rotatory radial movements of the forearm and hand. Moreover, in forms presenting the origin by two heads from the pectoral girdle, either or both of the girdle heads may be connected with either, or, by division of the insertion, with both of the forearm bones.

Consequently the analysis of the Biceps muscle would result as follows, retaining the names "glenoid" and "coracoid" as designating respectively the outer and inner girdle origins of the muscle :

Complete type form of muscle, four heads.

1. Gleno-radial.	*3. Coraco-radial.*
2. Gleno-ulnar.	*4. Coraco-ulnar.*

This type becomes modified in various forms by reduction and elimination of two or more heads, so as to present a number of variations. Krause* first pointed out this quadricipital character of the human Biceps, based on careful dissection of the muscle and analysis of the fibres at the insertion. He also collected a number of comparative anatomical facts in support of this view of the compound character of the muscle. A by no means exhaustive consideration of the structure of the Biceps in the lower animals reveals the existence of the following reduc-

---

* W. Krause, Specielle und Macroscopische Anatomie, Hanover, 1879, p. 223.

tions and combinations of the four type heads of origin and
their insertion.

I. *Gleno-radial alone.*

A typical instance is presented by Talpa europea. Geohegan * describes the muscle in this animal as follows: The Biceps is single-headed, arising just above the glenoid articular
surface by a long tendon which passes through a groove in the
extreme anterior end of the humerus; this groove is at first a
tunnel. It is only after it has passed the elbow that the muscle
becomes fleshy. Insertion below the middle of the radius.
Welcker † corrects this statement to read that the insertion is
above the middle of the radius.

Other forms presenting the Glenro-radial muscle are: Nycti-
pithecus, Stenops, Horse, Ruminants,‡ Lutra vulgaris.§

II. *Coraco-radial alone.*

This form of the Biceps muscle is presented by Orycteropus
capensis, Rhinoceros, Frog, Toad, Lacerta, ‖ Phoca vitulina.¶
In this animal the muscle arises from the coracoid process,
is inserted into the radial tuberosity, and is combined with a
short head, which is connected with the Brachialis anticus.

III. *Gleno-ulnar alone.*

The muscle occurs in this form in Hyrax capensis and in Ro-
dents generally.**

IV. *Coraco-ulnar alone.*

Humphry,†† in describing the muscular system of Cryptobranchus japonicus gives the following account of the Coraco-
brachialis longus.

"This is the largest muscle arising from the coracoid. It
arises from the hinder edge of the coracoid and divides into two
portions. Of these the larger and inner or lower division is inserted into the ulnar edge of the humerus for a quarter of an
inch above the internal condyle; the other division, being nearly

* Geohegan, Myology of the Fore-limb of Talpa europea, Proceedings Dublin
Biolog. Assoc., Vol. I., 1875, p. 5.
† Welcker, H., Archiv f. Anatomie und Entwicklungsgeschichte, 1878, p. 23.
‡ Krause, op. cit., p. 223.
§ Lucae, J. C. G., Die Robbe und Otter (Phoca vitulina und Lutra vulgaris) in ihrem
Knochen und Muskelskelet, Frankfurt a/M, 1873, p. 204.
‖ Krause, op. cit., p. 223.
¶ Lucae, op. cit., p. 199.
** Krause, op. cit., p. 223. W. Krause, Anatomie des Kaninchens, p. 107.
†† Op. cit., p. 83.

as large, is partly inserted into the side of the long tendon of the Biceps, while a bundle of its fibres is continued on over the elbow, and is inserted into the ulna near the joint. The last described division must represent the short or coracoid origin of the Biceps in man. There is no trace of it in Menobranch, Axolotl or Newt. The muscle in them, though large, is confined to the humerus in its insertion."

The Coraco-radialis or Biceps arises from the external surface of the coracoid, between the Epicoraco-brachial (Pectoralis minor) and the short Coraco-brachial, as a fan shaped muscle, the fibres of which pass across the short Coraco-brachial and soon converge into a long tendon, which runs down beneath the Pectoral. Having passed the Pectoral it receives the fibres of the long Coraco-brachial, passes over the elbow joint, and is inserted into the palmar surface of the upper end of the radius close to the joint. It is supplied by the nerve which perforates the scapula and which supplies also the superficial Coraco-brachial.

V. *Gleno-radial and Coraco-radial.*

Meckel * describes the double-headed Biceps of Ornithorrhynchus paradoxurus, one head arising from the anterior, the other from the posterior coracoid, the muscle passing to be inserted into the radius.

VI. *Gleno-radial and Gleno-ulnar.*

Dog.

Origin: By a single strong tendon from the edge of the glenoid fossa, the tendon passing through the capsular ligament of the joint.

Insertion: By a strong tendon chiefly into the ulna, although attached also to the radius by a smaller slip.

Other forms presenting the same arrangement of the muscle :

Cholœpus didactylus.†

Two-toed sloth.

Origin: By a long and strong tendon from the glenoid border of the scapula, passing down through the intertubercular groove in the capsule of the shoulder. The muscle descends on the arm, dividing into two strongly developed bellies, one passing with the Pectoralis and Deltoid to the tuberosity of the radius, the other with the Brachialis anticus to the coracoid process of the ulna.

---

* System der Vergleichenden Anatomie, p. 516.

† Lucae, J. C. G., Der Fuchsaffe und das Faulthier (Lemur macaco and Choloepus didactylus) in ihrem Knochen-und Muskelskelet, Frankfurt a/M, 1882.

The combination of gleno-radial and gleno-ulnar muscle is also noted in the Pig and Cat.

VII.   *Gleno-radial and Coraco-ulnar.*

Uromastix spinipes.*

Origin: by two distinct portions from coracoid, one tendinons from proximal part, the other muscular more laterally from anterior part, corresponding with the coracoid and glenoid origins in man.

Insertion: After being joined by Brachialis anticus, which is large, arising from the lower part of the humerus, the conjoined tendon is inserted into the radius and ulna.

VIII.   *Coraco-radial and Gleno-ulnar.*

In Marsupials, in which the two muscles are entirely separate from each other.†

IX.   *Coraco-Radial and Coraco-Ulnar.*

This arrangement of the muscle is found in Emys and Chameleon.‡

A general consideration of the facts above adduced leads to the natural conclusion that the Biceps muscle of the fore-limb, or its homologue, presents throughout the vertebrate series evidences in its structure and arrangement of a general type-plan of construction, modified in the different groups by the functional requirements of the limb, and in all probability by the varying relations assumed by the tendon of origin to the capsule and the cavity of the humero-scapular articulation.   I believe that this fundamental type-construction of the appendicular muscular system is the element to which we must refer for the explanation of deviations from the arrangement normally found in any one species.   Partial returns to the potential type-form constitute reversions of far broader significance than those usually grouped under the head of atavism.

In respect to the human subject, as well as in the case of the remaining vertebrates, and especially mammalia, our exact knowledge regarding the lines of descent of present orders and sub-orders is still so rudimentary and imperfect, that the impropriety of referring nearly all muscular variations to atavism and direct inheritance becomes at once apparent.   The fact that

*Humphry, op. cit., p. 64.
†Krause, op. cit., p. 223.
‡Krause, op. cit., p. 223.

in case of any given human muscular variation, a muscle of similar character is found in one of the lower vertebrates does not warrant the assumption that both are derived by inheritance from an immediately precedent common ancestral form. The form in which the variant human muscle appears normally may be incalculably far removed from man, may even belong to a different vertebrate class. That the structural coincidence of the two muscles is to be taken as indicating anything more than the most generalized relationship of vertebrates is difficult to believe. For many of the aberrant muscular conditions observed in man a very comprehensive view as to their derivation must be adopted. I believe that we are right in referring such variations, as will be considered in detail below, to the development of an inherent constructive type, abnormal for the species in question, but revealing its morphological significance and value by appearing as the normal condition in other vertebrates.

The question, as far as it affects the variations to be considered, may be represented graphically somewhat in the following manner :

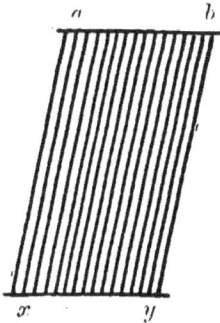

FIG. 1. Type form of Muscle.   FIG. 2. Cleavage variation.

In Fig. 1 let the line *a–b* represent the skeletal origin, the line *x–y* the corresponding insertion of a muscular plane. Considering this arrangement as the type, in which the entire space between origin and insertion is occupied by an uninterrupted muscular plane, it will become apparent that modifications of this type can take place in two ways.

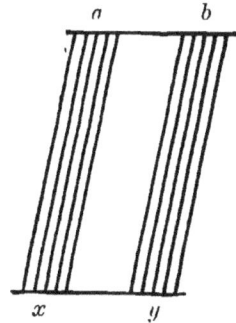

1. Cleavage variations, retaining in general the original scope of origin and insertion, in which the original muscular plane appears as two or more distinct muscles. (Fig. 2.)

2. Reduction variations, where a portion of the origin or of

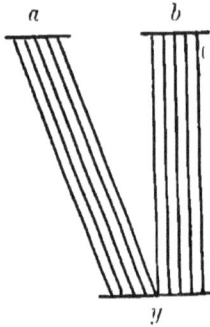

FIG. 3. Reduction variation; Origin retained; Insertion reduced.

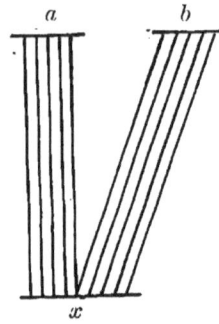

FIG. 4. Reduction variation; Origin retained; Insertion reduced.

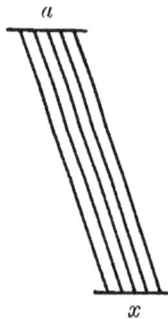

FIG. 5. Reduction variation; Origin and Insertion reduced; Points retained.

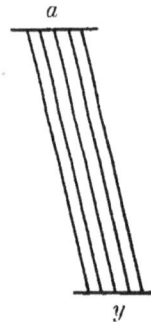

FIG. 6. Reduction variation; Origin and Insertion reduced; Points exchanged.

FIG. 7. Reduction variation; Origin and Insertion reduced; Points retained.

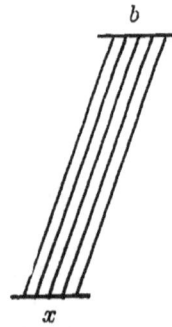

FIG. 8. Reduction variation; Origin and Insertion reduced; Points exchanged.

b   *a*

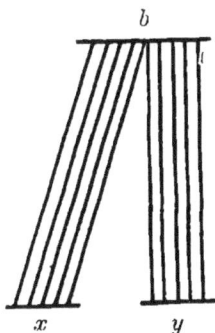

FIG. 9. Reduction variation; Origin   FIG. 10. Reduction variation; Origin
reduced; Insertion retained.   reduced; Insertion retained.

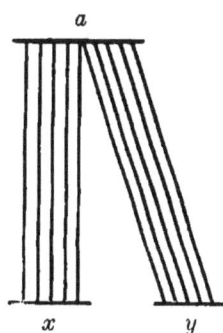

the insertion, or of both, is eliminated, necessitating, under some conditions, a change in the direction of the muscular fibres.

These variations are represented schematically by Figs. 3–10, it being of course evident that the arbitrarily selected points *a* and *b*, and *x* and *y*, may be placed anywhere along the lines of origin and insertion respectively.

In defining the following variations in the human Biceps I prefer to accentuate this relation to a fundamental vertebrate type-plan, and, at the expense of introducing an additional term into the complex reversional nomenclature, I will speak of them as " *Myo-typical Reversions.*"

In man the Biceps is composed of the Gleno-radial and Coraco-radial divisions, combined with a superficial ulnar fasciculus, possessing a fascial insertion along the ulnar border of the fore-arm by means of the semilunar fascia.

We may group the variations which concern us here with reference to the derivation and destination of the accessory portions. representing additional glenoid and coracoid heads, which have lost their distal ulnar attachment.

I. Gleno-ulnar Head.

Appears in the following forms :

1. Capsulo-pectoral Tendon.

Diagnosis : Tendinous fibres in the form of scattered bundles, or more compactly arranged as distinct tendon-bands, arising in conjunction with the capsular ligament from the glenoid margin of the scapula, differentiating from the capsule at the upper border of the intertubercular groove and extending downwards over the long gleno-radial head, roofing in the bicipital canal and merging with the deep surface of the tendon of the Pectoralis major, or, in some instances, extending beyond the pectoral tendon to the deep fascia of the arm.

This is the most common form in which the variation presents itself.

Cases :

1. ♀, white, U. S., aet. 62.   January 30, 1894.

Plate XVII.   Right upper extremity :

A distinct tendon, imbedded in the capsule of the joint, overlying the intertubercular groove, covering the long bicipital tendon and passing to the deep surface of the Pectoralis major tendon, on which it spreads out in the upper third, terminating by interlacing with the pectoral tendon fibres.

Left upper extremity :

Presents the same slip, more strongly developed, extending nearly to the lower border of the Pectoralis major tendon.

Plate XVIII. Capsule between the tuberosities partly divided, to show the deep position of the tendon imbedded in the shoulder capsule, and indicating a tendency toward intra-articular immigration.

Plate XIX. The same joint opened from behind, with head of humerus removed. The thickened strand on the inner surface of the anterior wall of the capsule, just in front of the long Biceps tendon, is directly continuous with the fibres of the tendon slip.

2. ♂, Austria, aet. 65.   October 11, 1893.

Plate XX.   Right upper extremity.

Strong fibrous band arising from capsule over lesser tuberosity, and descending over long tendon of Biceps, connected with deep surface of Pectoralis major tendon, to the lower margin of which it extends.

A tendon band from anterior part of capsule passes to outer margin of short head, indicating tendency to subdivision of this head.

3. ♀. Germany, aet. 84.   November 9, 1893.

Plate XXI. Right upper extremity.

Well marked strong fibrous fasciculus, extending from capsule of shoulder joint over the bicipital groove and contents to the deep surface of the Pectoralis major tendon. Some fibres merge with the pectoral tendon, while others continue below its inferior margin to join the deep fascia of the arm.

The left upper extremity of the same subject presents a transition to the form next to be described (Gleno-ulnar muscle). A strong tendinous band, incorporated at its origin in the capsule of the shoulder joint, passes downward, covering in the bicipital groove and attaching itself to the upper margin and deep surface of the Pectoralis major tendon. In part continuous with this tendon band, in part arising independently from the deep surface of Pectoralis tendon, four muscular slips arise, the two upper passing backward, the two lower downward, to incorporate themselves with the substance of the Biceps muscle. The bundles are all distinctly muscular in character.

The following additional instances, in which the above described band was present, were noted in the series :

4. ♂, U. S. white, aet. 38. December 27, 1893.
Small glenoid tendon slip joining deep surface of Pectoralis major tendon over the bicipital groove. Present on both sides.

5. ♀ Ireland, aet. 72. January 9, 1894.
Slip present on both sides.

6. ♂ Ireland, aet. 40. December 21, 1893.
Right upper extremity.
Large tendon band, arising from capsule external to long head of Biceps, inserted on deep surface of Pectoralis major tendon.

7. ♂ Ireland, aet. 32. November 28, 1893.
Right side.

8. ♂ Germany, aet. 54.
Large band on left side. Insertion confined to Pectoralis major tendon.

9. ♂ Germany, aet. 52. December 14, 1893.
Right side.
Very large tendon ½ inch wide, passing from glenoid margin interwoven with the capsule, to the united bellies of Biceps, with intermediate attachment to the deep surface of Pectoralis major tendon.

10. ♂ Ireland, aet. 40.
Both extremities.
Strong tendon from glenoid edge to deep surface of Pectoralis major tendon.

11. ♂ Italy, aet. 50.
Right side.
Very broad tendon slip, in part continued beyond Pectoralis tendon into the Biceps.

12. ♀, U. S., negro, aet. 24.
Left side :
Glenoid tendon to junction of heads of Biceps.

2. GLENO-ULNAR MUSCLE.

Occurs in three forms or variations :

(a.) *Origin identical with that of the Capsulo-pectoral tendon, passing downward and presenting a more or less close connection with the deep surface of the Pectoralis major tendon. Becoming muscular a short distance beyond the lower border of the Pectoralis tendon the slip passes obliquely inwards across the long head, fusing with its anterior surface and inner margin, or joining the outer margin of the short coracoid head.*

Cases :

1. ♂, U. S., white, aet. 24. November 29, 1893. Plate XXII. Right side.

A strong shining tendinous band arises from the glenoid margin incorporated with the shoulder capsule. Passing down and emerging from beneath the edge of the coraco-acromial ligament, the tendon becomes closely connected with the deep surface of the Pectoralis major tendon, interlacing with its fibres at right angles, lying upon the long tendon of the Biceps and forming the main portion of the roof of the inter-tubercular groove. Toward the lower portion of the pectoral tendon the inner half of the slip gradually separates itself from the same and forms a small muscular belly which joins the outer margin of the short biciptal head near its junction with the long head. The outer portion of the tendon slip gradually passes deeper into the substance of the Pectoralis tendon, with which it is intricately interlaced. On the bottom of the inter-tubercular groove are several well marked longitudinal tendinous bands, connected below with the tendon of the Pectoralis major ; above they in part separate from the tendon, passing upward and inward and crossing at right angles the tendon fibres of Latissimus dorsi and Teres major ; in part they go upward and outward, lying in the floor of the bicipital groove under cover of the long Biceps tendon.

A small third internal humeral head of the Biceps is present in the same arm, derived from the Coraco-brachialis insertion.

2. ♂, U. S., white, aet. 50. Plate XXIII. Right upper extremity.

Tendon slip with intermediate attachment to deep surface of Pectoralis major tendon, arising in connection with capsular ligament from the glenoid margin, and passing to the inner border of the long bicipital head, at the line of junction with the short head.

3. ♂, Ireland, aet. 32. November 29, 1893. Plate XXVIII. Left upper extremity of subject, whose right upper extremity presents the variation described under " b 2," Plate XXVII. *vidi ibid.*

4. ♂, U. S., negro, aet. 50. December 14, 1893. Plate XXIV. Right side.

Tendon in part fused with deep surface of Pectoralis major

tendon at the lower border of which a short muscular belly develops which joins the Biceps.

Between the long and short bicipital heads an additional tendon slip arises from the capsule, near the base of the coracoid
process, and, passing down on the outer side of the short Biceps
tendon, receives some upper fibres of insertion of a short Coraco-
brachialis superior, and then joins the outer margin of the short
Biceps head just before the latter meets the long head.

The left side of the same subject presents an instance of the
next succeeding form.

·5. ♂, Germany, aet. 52.   December 14, 1893.

Very large accessory glenoid head, $\frac{1}{2}''$ broad, passing to the
united bellies of the Biceps.   Upper tendinous portion very
distinct and adherent to Pectoralis tendon.

6. ♀, U. S. white. aet. 23.   January 30, 1894.

Plate XXV.   Left upper.

Gleno-ulnar tendon connected but loosely with deep surface
of Pectoralis major; leaves tendon of Pectoral as a distinct
band at about the middle of its posterior surface and a short
distance below the distal margin of the pectoral tendon, expands into a superficial broad and flat muscular belly, which
descends upon the long bicipital head, lying in close connection
with the outer margin of the coracoid head from which it is
completely separable to a point midway between the lower
border of the Teres major and Latissimus dorsi and the elbow.
The fibres of this accessory head then join the coracoid head
along its external border.   About 3 cm. above the internal
condyle the long head joins the tendon of insertion, developed
on the deep and external aspect of the combined coracoid and
accessory heads.

The bicipital fascia receives some superficial and oblique
fibres (decussating) from the outer and anterior aspect of the
radial tendon.

The right arm of the same subject presents a Gleno-epitrochlear slip (*vide infra*).

(*b.*) *Tendon of origin of Gleno-ulnar muscle is completely
free from Pectoralis major tendon, under cover of which it lies.*

Cases :

1. ♂, U. S., negro, aet. 50.   December 14, 1893.

Plate XXVI.   Left upper extremity of subject whose right
extremity is described under " a 4," Plate XXIV.

Long, distinct glenoid tendon, overlying long bicipital tendon, under cover of Pectoralis major, but not connected with
the same.   Becomes muscular opposite lower third of pectoral
tendon and crosses over long bicipital head to join its inner
margin at about the middle of the arm.

2. ♂, Ireland, aet. 32. November 29, 1893.

Plate XXVII. Right upper extremity. Under cover of insertion of Pectoralis major, a broad tendinous band descends over the intertubercular groove, derived from the shoulder capsule and covering the long tendon of the Biceps, to which it is connected by a thin but strong tendinous lamina. The lower portion of this band terminates in a broad muscular sheet, which turns inwards to join the medial margin of the long head.

A few scattered tendinous fibres descend obliquely from the capsule to the outer margin and anterior surface of the lesser head.

The left upper extremity of the same subject presents the gleno-ulnar slip with intermediate connection to the deep pectoral tendon. (Variety "a.") Plate XXVIII.

Here a strong tendinous band, arising from the intertubucular portion of the capsule, and connected with the upper part of the deep surface of the Pectoralis tendon, becoming free below, crosses obliquely over the long bicipital tendon to fuse with the inner margin of its muscular belly a short distance before the same is joined by the coracoid head.

A small tendinous bundle (Capsulo-intermediate fibres), arising from the coracoid portion of the capsule, accedes to the outer part of the coracoid head.

Insertion of Biceps in this arm: Radial tendon derived as usual from the deep surface of the combined muscle. Superficially some fibres pass from the outer part of the muscle across the radial tendon into the semilunar fascia, which is inserted by a forked process into the radial as well as the ulnar side of the deep forearm fascia.

c. *Transition forms between "a" and "b," and variations.*

Cases :

1. ♂, Ireland, aet. 53. March 16, 1894.

Plate XXIX. Left upper extremity.

Gleno-ulnar tendon slip, presenting a reduplication of the origin. The outer band is connected with the Pectoralis major tendon ; the inner is free. They join opposite the middle of the pectoral tendon and develop a stout, muscular belly, which can be followed as far as the elbow, where a few of the deeper fibres pass into the inner margin of the radial tendon of insertion, whereas the remainder of the deep and all the superficial fibres of the accessory head, joined by some from the outer margin of the main muscle, pass inward into the bicipital fascia.

2. ♂, Ireland, aet. 35. March 14, 1894.

Plate XXX. Left upper extremity.

Gleno-ulnar tendon, double at origin, a portion passing into

the pectoral tendon, the remainder into the outer margin of the
short bicipital head and into the Coraco-brachialis, which muscle
is strongly developed in its upper part.
3. ♂, Germany. aet. 64. October 14, 1893.
Plate XXXI. Right upper extremity.
Double glenoid tendon, the outer division giving off a branch
to join the deep surface of the pectoral tendon. The tendons of
origin then unite and form a muscular belly which passes down
and in over the long bicipital head to join the deep surface and
outer margin of the short head.

3. GLENO-EPITROCHLEAR TENDON.

Diagnosis: Origin as Capsulo-pectoral fibres from glenoid
margin with capsule of shoulder. Connected with deep surface
of Pectoralis major tendon at the lower border of which the
fibres are gathered into a long, slender tendon which passes
obliquely down and in, across the Biceps, brachial vessels and
large nerves, to be inserted into the anterior part of the internal
epicondyle.

The structure exhibited by this variation in its upper portion
is such as to admit of no doubt of its identity with the fibres
described above as Capsulo-pectoral and Gleno-ulnar with intermediate pectoral connection.

The lower long tendon with its epicondylar implantation represents, I believe, a rudimentary Gleno-ulnar muscle, whose insertion has been shifted to the internal epicondyle. The latter
is a point frequently selected for the insertion of aberrant and
rudimentary tendon slips, as instanced by the Dorso-epitroch-
learis and Chondro-epitrochlearis muscles.

The supposition that the variation just mentioned and some
of the previously described bicipital variations can be brought
into connection with aberrant pectoral muscle or tendon slips
is refuted by the constant connection with the capsule, and by
means of the same with the glenoid margin of the scapula.

The pectoral varieties described as Chondro-humeral, Chondro-
coracoid and Chondro-epitrochlear muscles are connected at
times with the lower border of the pectoral muscle and tendon,
but they are evidently derivatives from the pectoral plane, and
never assume the characteristic relation to the deep surface of
the pectoral tendon and shoulder capsule exhibited by the bicipital variations above described.

I have met with four well-marked instances of the Gleno-
epitrochlear tendon in the series examined.

Cases :

1. ♂, Germany, aet. 66. January 11, 1894.
Plate XXXII. Left upper extremity.

Glenoid tendon, overlying long Biceps tendon and fusing with the under surface of the Pectoralis major tendon in its broad upper part. About the middle of the pectoral tendon this connection ceases, and the tendon bundle becomes narrower and more sharply defined. It remains tendinous throughout and passes obliquely downward and inward over the Biceps, median nerve and brachial artery to be attached to the anterior aspect of the internal epicondyle.

In confirmation of the view expressed as to the nature of this variation the right upper extremity of the same subject presented the Capsulo-pectoral slip, as a well-marked tendon, connected with the deep surface of the Pectoralis major; also a separate slender tendon from the coraco-acromial ligament to the outer margin of the short bicipital head.

2. ♂, Ireland, aet. 42. November 13, 1894.
Plate XXXIII. Left upper extremity.

A tendon from the glenoid margin, interwoven with the capsule of the shoulder joint, passes over the long Biceps tendon, covered by the Pectoralis major tendon, to the deep surface of which it is partially attached, and continues beyond the pectoral tendon as a slender but very distinct band, passing obliquely downward and inward over the brachial artery and the large nerves to the internal epicondyle, blending with the lower quarter of Struther's ligament.

3. ♂, Germany, aet. 66. January 11, 1894.
Left upper extremity presents exactly the same arrangement.

4. ♂, U. S., white, aet. 23. January 30, 1894.
Right upper extremity.

Gleno-ulnar fibres, rather weak in upper portion, on posterior surface of Pectoralis major tendon.

From the lower part of the posterior surface of pectoral tendon, and from inferior border of deep reflected part, fibres are derived which form a slender tendon, passing down and in, superficial to and obliquely across Biceps and brachial vessels and nerves, to be inserted on the anterior surface of the internal epicondyle.

4. M. BRACHIO-ULANRIS LATERALIS.

I have placed under this designation certain of the variations of the Biceps, constituting accessory third humeral heads, arising from the anterior or external surface of the shaft, because a careful examination affords grounds for regarding them as distal portions of a Gleno-ulnar muscle, which has lost its proximal or origin portion, represented by the variations just described, and has obtained a secondary attachment to the humerus near the bicipital groove and the insertion of the Pectoralis major.

This group would therefore form the last of a series of variations of a muscle arising from the glenoid margin and finding its insertion at the ulnar border of the forearm.

Cases:

1. ♀, Ireland, aet. 27.   October 25, 1893.

Plate XXXIV. Both upper extremities present the same variation.

A third additional Biceps head arises from the outer surface of the humeral shaft and capsule of the shoulder joint, and from a strong tendinous band into which the upper fibres of a Coracobrachialis brevis are inserted.

The connections with the shoulder capsule, found in the Capsulo-pectoral, Gleno-ulnar and Gleno-epitrochlear variations, is retained in this instance as one of the points of origin of the accesssory head.

2. ♀, U. S. negro, aet. 24.   December 27, 1893.

Plate XXXV. Right upper extremity.

A third bicipital head arises from the anterior surface of the shaft of the humerus, almost directly continuous with the lower margin of the Pectoralis major tendon and attached to the lower portion of the bicipital groove.

The muscle passes down on the outer side of the arm, separated from the upper and outer portion of the Brachialis anticus by some branches of the musculo-cutaneous nerve and muscular arterial branches. It then turns inwards beneath the main muscle to join the common tendon on the deep aspect and along the ulnar margin.

5. COMBINATION OF CAPSULO-PECTORAL OR GLENO-ULNAR WITH GLENO-EPITROCHLEAR FIBRES.

Case:

♂, U. S., white, aet 23.   January 30, 1894.

The two extremities present the variations already described as follows:

Right upper.   Gleno-epitrochlear tendon (No. 4).

Left upper.   Gleno-ulnar muscle, "a," No. 6, Plate XXV.

6. COMBINATION OF TENDINOUS GLENO-ULNAR AND CORACO-EPI-TROCHLEAR.

Case:

♀, Ireland, aet. 72.   January 9, 1894.

Plate XXXVI. Right upper extremity.

A slightly marked tendinous gleno-ulnar slip passes from the capsule of the shoulder over the intertubercular groove, with intermediate attachment to deep surface of pectoral tendon, to the angle of union between the long and short bicipital heads.

From the anterior surface of the lower part of the Coraco-brachialis, near its insertion, is derived a slender muscle-slip, which, becoming tendinous, passes obliquely down and in over brachial vessels and median nerve to be attached to the internal epicondyle.

The left upper extremity of the same subject presents a small tendinous band passing from the capsule to the deep surface of the Pectoralis major tendon.

II. CORACO-ULNAR HEAD.

This muscle is found usually very closely fused in its proximal part with the short or corao-radial head of the Biceps, and with the Coraco brachialis. But its distal portion appears frequently as the " *third internal humeral head.* "

The following variations are to be noted :

1. CAPSULA-INTERMEDIATE TENDON.

Origin: Tendon fibres derived from the capsule of the shoulder joint between the coracoid process and bicipital groove. The slip passes down to the angle of fusion between the long and short heads of the Biceps.

Cases :

In combination with capsulo-pectoral fibres in the following previously described instances :

1. ♂, U. S., negro, aet. 50. December 14, 1893.
" 2ª," No. 4. Plate XXIV. Well marked on right side.
2. ♂, Ireland, act. 32. November 29, 1893.
"2ᵇ," No. 2. Plate XXVII. Especially well marked on left side.
3. ♂, Germany, aet. 66. January 11, 1894.
Right upper extremity. " 3 " No. 1.
Additional cases.
4. ♀, Italy, aet. 30. October 31, 1893.
Both upper extremities.

A strong fibro-tendinous band arising from the inner part of the capsule of the shoulder passes vertically down to be attached to the septum between the long and short Biceps heads.

5. ♂, U. S., white, aet. 27. December 31, 1893.
Right upper extremity.

Slender tendon from inner part of shoulder capsule to outer margin of lesser bicipital head.

Macalister * describes several varieties of additional coracoid heads of the Biceps (catalogue numbers 11–14 incl.). The accessory portion may join the main body of the muscle, or else it may unite with the normal coracoid head, before that portion of the Biceps joins the long head. Additional coracoid origins from the Coraco-acromial ligament and from the insertion tendon of the Pectoralis minor are also mentioned by the same author.

2. Coraco-epitrochlaris.

As in the case of the Gleno-ulnar head certain instances occur in which an additional coracoid head passes to the internal epicondyle. I have met this arrangement in two forms :

a. *Coraco-epitrochlear tendon.*

Cases :

1. ♂, U. S. white, aet. 47.   March, 1894.

Plate XXXVII.   Right upper extremity.

A slender, firm tendon arises from the coracoid process at the inner border of the short bicipital head and superficial to the Coraco-brachialis origin. It passes downward and slightly inward obliquely over the brachial artery and the large nerves, receives near the elbow an accession of fibres from the internal intermuscular septum, and is inserted into the internal epicondyle.

2. ♂, Ireland, aet. 63.   November 14, 1894.

Right upper extremity :

A thin tendinous slip arises from the intermuscular septum betwteen the short head of the Biceps and the Coraco-brachialis, passes downward and inward, over the musculo-cutaneous nerve' to the internal epicondyle.

The musculo-cutaneous nerve passes entirely below the Coraco-brachialis, between this muscle and the short bicipital head,' the former receiving its nerve higher up by a separate branch from the outer cord of the brachial plexus.

3. ♂, U. S. white, aet. 34.   February 1, 1894.

Left upper extremity :

A tendon slip arises from the coracoid process and tendon of origin of the Coraco-brachialis and short bicipital head ; becoming free about 2 cm. below the coracoid it passes downward and inward as a distinct tendon to be inserted into the internal epicondyle.

4. ♂, U. S. white, aet. 39.   December 14, 1893.

Left upper extremity :

A tendon slip from origin of Coraco-brachialis and short head

---

* Op. cit., p. 80.

of Biceps crosses over the brachial vessels and the nerves to the internal epicondyle.

The long tendon of the Biceps in this arm is doubled.

5. ♀, Ireland, aet. 72.   January 9th, 1894.

Combined with tendinous Gleno-ulnar band in right arm (vide I. 6, Plate XXXVI).

    b. Coraco-epitrochlear and Gleno-epitrochlear tendon combined.

Case :

    ♂, U. S., white, aet. 46.   November 11, 1893.

Plate XXXVIII.   Right upper extremity.

A coraco-epitrochlear tendon, arising from Coracoid process at point of junction between Coraco-brachialis and short bicipital head, passes down as above described over the brachial vessels and the large nerves, and is joined at about the middle of the arm by a second similar tendon which arises from the outer portion of the capsule of the joint, descends beneath Pectoralis major tendon and then crosses obliquely down and in over the short bicipital muscle.   The conjoined tendons continue downward in the line of the Internal intermuscular Septum and are inserted as a single band into the internal epicondyle.

    c. M. Coraco-epitrochlearis.

Case :

    ♂, Germany, aet. 29.   January 11, 1894.

Plate XXXIX.   Left upper extremity.

A slender superficial tendon arising from the coracoid process develops a thin fusiform muscle at about the middle of the arm, which passes down to the internal epicondyle, lying upon the brachial vessels and the large nerves.

The obvious connection of the Coraco-epitrochlear variations first described with the Coraco-brachialis inferior will be considered in dealing with the relation of the Biceps to that muscle.

3. M. Brachio-ulnaris Medialis.

Under this head I have placed the variations which include a third bicipital head, arising from the inner surface. of the shaft of the humerus, either from the interval between the Coracobrachialis and Brachialis anticus. or directly from the latter muscle, or from the insertion of the Coraco-brachialis and continuous with that muscle.

It has seemed to me, in examining carefully this frequent variation, that we have to deal here with a Coraco-ulnar head which has lost its girdle attachment and has transferred its origin to the shaft of the humerus, modifying its insertion by joining the remainder of the Biceps muscle.

The relation of this variation to the Brachialis anticus and Coraco-brachialis is obvious and will be referred to below.

In the above series I have encountered this third internal head in the following instances :

1. ♂, England, aet. 73. January 11, 1894.
Left upper extremity.
Internal bicipital head derived directly from the humerus, external to and about 5 cm. above the Coraco-brachialis insertion, between it and the Brachialis anticus, completely free from the latter muscle.

The musculo-cutaneous nerve, after piercing the Coraco-brachialis, passes between the additional head and the main portion of the Biceps muscle.

2. ♂, Ireland, aet. 46. January 11, 1894.
Third internal bicipital head arising from the outer margin of the Coraco-brachialis insertion, and in close connection with the superior and internal origin of the Brachialis anticus.

3. ♂, Ireland, aet. 37. October 17, 1893.
Right upper extremity.
Third bicipital head arising from inner surface of shaft of humerus at the Coraco-brachialis insertion. Separated from the main Biceps muscle by the large muscular branches of the musculo-cutaneous nerve, which perforates the Coraco-brachialis.

The internal head has further been noted in the following cases :

4. ♂, Ireland, aet. 46. January 24, 1894.
Right upper extremity.
5. ♂, Ireland, aet. 62. November 28, 1893.
Right upper extremity.
6. ♂, Germany, aet. 28. October 5, 1893.
Left upper extremity.
Additional head, derived in this instance from the insertion of the Coraco-brachialis, goes mainly into the semilunar fascia.

7. ♀, Germany, aet. 84. November 9, 1893.
Left upper extremity.
8. ♀, U. S., white, aet. 68. November 22, 1893.
Right upper extremity.
Small third internal head, derived chiefly from the tendon of of the Coraco-brachialis, with which it is continuous.

9. ♂, Germany, aet. 60. December 14, 1893.
Right upper extremity.
10. ♂, Bohemia, aet. 30. December 19, 1893.
Left upper extremity.
11. ♂, Italy, aet. 40. December 14, 1893.
Right upper extremity.

II. RELATION OF BICEPS TO THE ADJOINING DEEP VENTRO-APPENDIC-
ULAR FIBRES OF THE CORACO-BRACHIALIS, AND, III. TO THE
DEEP INTRINSIC FLEXOR OF FOREARM, BRACHIALIS
ANTICUS. ·

The consideration of the variations, described above as the
Humero-ulnar internal head and the Coraco-epitrochlear slips, in-
dicate the intimate relation existing between the Biceps and the
Brachialis anticus and Coraco-brachialis.

Humphry,[*] in describing the muscles of the Cryptobranch,
accentuates the close relation of the Coraco-brachialis longus
and Biceps of this animal. He finds that the former muscle
divides into two portions, one of which is inserted into the
ulnar edge of the humerus; the other, being nearly as large, is
partly inserted into the side of the long tendon of the Biceps,
while a bundle of its fibres continues over the elbow and is inserted into the ulna near the joint. Humphry regards this latter portion as the representative of the short or coracoid bicipital origin in man.

We find in this instance on the one hand the direct union of
the Coraco-brachialis with the Biceps, and on the other insertion
of part of the muscle into the ulna.

Again the Brachialis anticus is in some forms (Pteropus)
found to be in direct continuity with the Coraco-humeral,[†] and
in the Scinc the Biceps derives two factors from the humerus,
which occupy the position of the Brachialis anticus and are so
named by Rüdinger. Humphry sums the mutual relations of
these three muscles up as follows :

"They show the Biceps to be an intermediate between the
Coraco-humerals and Brachialis anticus, continuous with either
or both, and uniting them into one group, extending from the
coracoid, along the ulnar and palmar surface of the humerus,
to the radius and ulna."

I believe that we may properly regard the variations of the
Biceps above referred to in this light. Both the Coraco-epitrochlear slips and the Internal humeral heads speak for the
original unity of a muscular plane extending between coracoid
and ulna.

The separation of the radius as the rotatory element of the
forearm and hand, and the assignment of the corresponding
muscular function to the Biceps, have caused the elimination of
the ulnar segment of the muscle, leaving the Brachialis anticus
as the deep intrinsic flexor connected with the ulna, and reduc-

---

[*] Op. cit., p. 33.

[†] Humphry, op. cit., p. 164.

ing the Coraco-brachialis to a deep ventro-appendicular muscle confined in its insertion to the humerus.

It is only in the Third Internal humeral head of the Biceps, and in the Coraco-epitrochlear slips, that we still find the evidence of the original connection between these muscles, and see the reversion of the Biceps toward its lost ulnar segment.

It is only necessary to refer in this connection to the interesting account of the structure of the Coraco-brachialis given by Prof. Wood,* and to point out the significance of the occasional Coraco-brachialis longus.

Analysis of cases of Quadriceps flexor cubiti and evidences of the ulnar tendency of the Biceps in variations of the insertion.

In the above series five examples of a four-headed muscle have been encountered.

1. ♂, Ireland, act. 45.   January 9, 1894.

Plate XL.   Right upper extremity.

A third additional head (gleno-ulnar) arises by a flat tendon from the capsule of the shoulder-joint, between the long and coracoid heads.   It is separated above from the main muscle by the muscuto-cutaneous nerve, after the latter has perforated the Coraco-brachialis.   A fourth additional head (coraco-ulnar) arises from the inner surface of the shaft of the humerus, almost directly continuous with the Coraco-brachialis at its insertion, and separated from the Brachialis anticus by a branch to the latter muscle from the musculo-cutaneous nerve.

The left upper extremity of the same subject presents the additional internal humeral head from the Coraco-brachial insertion.

The case affords an example of the typical composition of the four-headed muscle.   The gleno-ulnar head passes to the deep and ulnar surface of the main muscle, which is joined lower down by the coraco-ulnar, the continuity of the latter with the Coraco-brachialis being well marked.   The superficial part of the muscle is constituted by the large gleno- and coraco-radial heads.

2. ♀, Ireland, aet 40.   October 25, 1894.

Plate XLI.   Both upper extremities present the same arrangement.

A third anomalous head arises from the glenoid margin and capsule of the shoulder, passes down over the bicipital groove, covering the tendon of the long head, supplied by a branch from the musculo-cutaneous nerve, after the same has perforated the Coraco-brachialis.   A fourth additional head arises from the humerus, along the outer margin of the tendinous Coraco-brachialis insertion.

The main portion of the musculo-cutaneous nerve, after per-

---

* Journal of Anatomy and Physiology, Vol. I., p. 44, 1867.

forating the Coraco-brachialis, passes first between the 3d and 4th additional heads, and subsequently turns down to lie between the accessory heads, a short distance above their junction with the tendon of insertion, and the Brachialis anticus.

The semilunar fascia is well developed, scattered fibres extending up over the brachial artery as high as the internal epicondyle.

This case only differs from the preceding one in the separation of the fourth accessory head from the Coraco-brachialis.

3. ♂, Germany, aet. 62. November 29, 1894.

Plate XLII. Right upper extremity.

A combination of an additional gleno-ulnar head with intermediate pectoral tendon attachment, and a fourth internal humeral head, arising between the Coraco-brachialis and Brachialis anticus.

The insertion is peculiar. The radial tendon is formed by the long head (gleno-radial) and by the deep portion of the coracoid and additional internal humeral heads (coraco-radial). The remaining superficial portion of the regular coracoid and of the internal humeral head (coraco-ulnar) is joined by the entire additional glenoid muscle (gleno-ulnar) and passes superficially inward into a strong tendinous semilunar fascia which is well separated from the radial tendon.

4. ♂, U. S. white, aet. 63. November 7, 1894.

Plate XLIII. Right upper extremity.

The gleno-radial and gleno-ulnar heads are well defined at their origin, fusing before meeting the coracoid segment. The 4th head is derived from the outer margin of the regular coracoid head (coraco-radial?), 5 cm. below level of lesser tuberosity, as a slender slip, about 10 cm. long, which joins the inner margin of the glenoid portion, before the latter fuses with the main coracoid muscle.

5. ♀, U. S. white, aet. 26. November 28, 1894.

Plate XLIV. Left upper extremity.

A third internal humeral head arises from the shaft of the humerus at the Coraco-brachialis insertion and joins the regular coracoid head along its ulnar margin, 2.5 cm. above the level of the elbow. (Coraco-radial and Coraco-ulnar).

The fourth head (gleno-ulnar) is derived from the long tendon, along its outer margin, under cover of the Pectoralis, by a tendon which becomes muscular at the lower border of the pectoral tendon and fuses about the middle of the arm with the external and anterior part of the Brachialis anticus.

Macalister * has found a similar slip once.

---

* Op. cit., p. 83.

The variation is interesting as affording an instance of continuity of the Biceps with the Brachialis anticus, and hence of
a direct ulnar destination of some of the bicipital glenoid fibres.
Variations of insertion pointing to ulnar attachment of Biceps.
1. ♂, Assyria, aet. 28.   November 15, 1894.
Plate XLV.   Right upper extremity.

A muscular belly, entirely separate from Brachialis anticus,
arises from the lower part of the inner margin of the humeral
shaft, and passes downwards and outwards, underneath the Biceps, dividing into two portions.   The internal division passes
to reinforce the semilunar fascia ; the external stronger bundle
dips into the cubital fossa to the inner side of the radial Biceps
tendon, joining it, and giving some fibres to the fascia of the
Supinator brevis.   The additional muscle in this instance is evidently a compound of portions of both Coraco-radial and Coraco-
ulnar, whose origin has shifted downwards to the humeral
shaft, the proximal portion remaining as the coracoid head of
the regular Biceps and the Coraco-brachialis.

The compound character of both the external radial tendon of
the Biceps and of the semilunar fascia (internal or ulnar tendon)
is well shown by this case.

2. ♂, Ireland, aet. 57.   October 7, 1893.
Right upper.
Tendon slip from radial biceps insertion into lesser head of
Pronator teres.

3. ♂, Germany, aet. 58.   January 23, 1894.
Right upper.
Tendon slip from radial Biceps tendon into Pronator teres and
deep fascia of forearm. ·

Macalister * describes insertions of the Biceps into the coronoid process, Pronator teres, Coronoid insertion of Brachialis
anticus, capsule of elbow and the origin of some of the flexor
muscles.

While completing this paper a number of additional bicipital
variations, bearing out the views expressed relative to the composition of the muscle, have come under observation.   Among
them the two following instances illustrate some of the important morphological features of the muscle so well that they are
added to this paper.

1. ♂, Ireland, aet. 67.   January 22, 1895.
Plate XLVI.   Left upper extremity.
I. Glenoid Heads.
The outer bicipital head (Gleno-radial) is well developed.   In
its upper portion the tendon is overlapped by a distinct band,

---

* Op. cit., p. 83.

arising from the capsule of the shoulder joint and attached to
the deep surface of the Pectoralis major tendon (Capsulo-pectoral variety of Gleno-ulnar head). About 4 cm. above the
elbow joint a slip separates from the inner border and deep surface of the main muscle, and, becoming tendinous, passes downward and inward to the anterior border of the ulna, just below
the coracoid insertion of the Brachialis anticus. (Persistent distal portion of Gleno-ulnar division).

II. Coracoid Heads.

A broad muscle, arising with the Coraco-brachialis from the
coracoid process, divides at the level of the lower Pectoralis
margin into : (a) The usual short or coracoid head, which passes
down and out and joins the outer head a little above the middle
of the arm (*Coraco-radial division*). (b). An internal muscle,
completely free from Coraco-brachialis, which immediately sub-
divides into two equal portions. Of these the posterior and internal muscle descends vertically over the Coraco-brachialis insertion, and terminates at the junction of middle and lower one-
third of the arm in a strong tendon, which passes down, at first
free and subsequently fused with Struther's ligament, to the internal epicondyle. (*Coraco-epitrochlear variety of Gleno-ulnar
head.*

The anterior muscle, lying upon the Brachialis anticus and the
internal intermuscular septum, remains completely free from
surrounding structures and terminates in a strong tendon which
passes over the elbow joint and is inserted into the coronoid
process of the ulna, just internal to the Brachialis anticus insertion. (*Typical Coraco-ulnar head of muscle.*)

This case is especially important, as it presents not only the
more common proximal vestiges of the obsolete ulnar divisions,
but also exhibits perfectly the rare distal or insertion portions,
in their complete form, attached to the ulna.

The separation of the Biceps insertion from the ulna and the
assignment of the muscle, as a supinator, to the radius, would
lead us to expect this disproportion, as regards frequency of occurrence, between reversions of the complete distal and proximal segments of the lost ulnar division. The distal or insertion portion of the ulnar division was the first to disappear at
the insertion into the ulna, and consequently reverts in a very
much smaller percentage in its complete form than the proximal or origin portion, whose existence has, so to speak, been
prolonged by the opportunity of uniting with the radial division.

The above instance exhibits these features of the muscle-plan
perfectly, and the preparation has been added to the Variation-
series of the Morphological Museum of Columbia College.

2. The second case, recently observed, which presents points of especial interest in connection with the subject of this paper, exhibits one of the important relations between Biceps and Brachialis anticus, and emphasizes the significance of the semilunar fascia, as representing the remains of an ulnar bicipital division which has lost its skeletal attachment, in accordance with the functional specialization of the muscle as the main supinator of the limb.

♀, Ireland, aet. 54.   March 15, 1895.

Plate XLVII.   Right upper extremity :

This case affords a well-marked example of the original connection between Biceps and Brachialis anticus.

The origin of the Biceps in this arm is normal, as is the arrangement of Coraco-brachialis. A strong muscular bundle separates from Brachialis anticus a short distance below the Coraco-brachial insertion. The outer and larger portion of this muscle joins the deep surface of the Biceps and passes with it to the radial insertion. The inner part continues downward and inward, gives off a narrow tendon which passes with the remainder of Brachialis anticus to the coronoid process, and then expands into the semilunar fascia, which is well developed, crossing obliquely over the brachial artery.

The additional muscle in this instance is evidently an Internal Brachio-ulnar muscle, which, however, presents not only the usual connection with the radius by means of the bicipital junction, but preserves its original ulnar insertion both by the tendon slip to the coronoid process and by the development of the entire semilunar fascia. The significence of the latter structure, entitling it to be considered as the distal portion of an ulnar bicipital segment which has lost its skeletal attachment, is strongly emphasized by the arrangement of the aberrant muscle in this subject.

PLATE XVIII.

Right upper extremity.
♀, white, U. S. aet. 62.
Gleno-ulnar head ; var. 1, Capsulo-pectoral tendon.

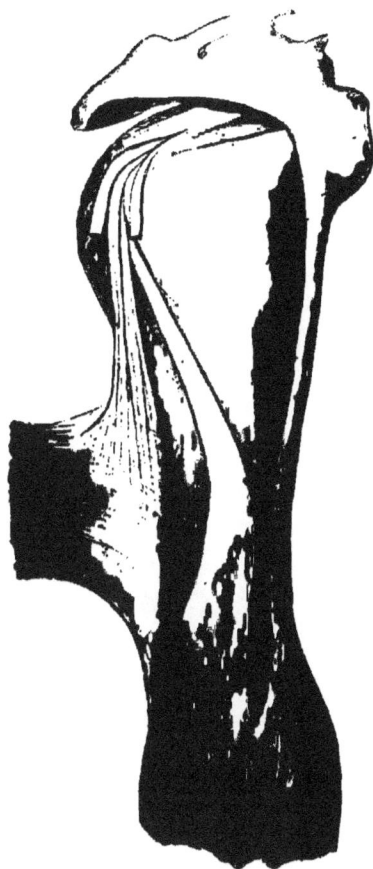

PLATE XIX.

Right shoulder joint of same subject, opened from behind, head of humorus removed, showing thickening of anterior capsule wall by the Capsulo-pectoral tendon.

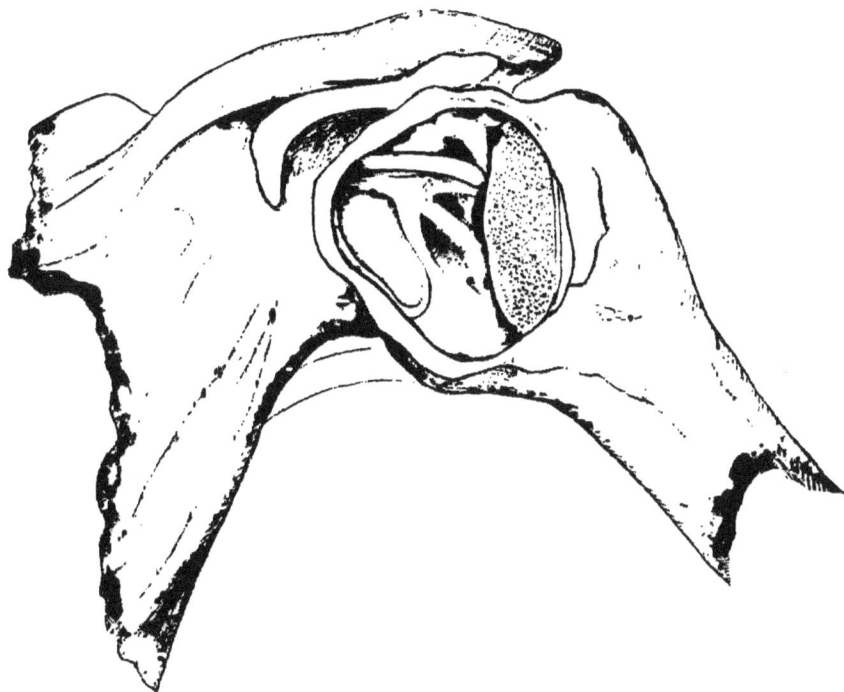

PLATE XX.

Right upper extremity.

♂, Austria, aet. 65.

Gleno-ulnar head ; var. 1, Capsulo-pectoral tendon.

PLATE XXI.

Right upper extremity.
♀, Germany, aet. 84.
Gleno-ulnar head, var. 1, Capsulo-pectoral tendon.

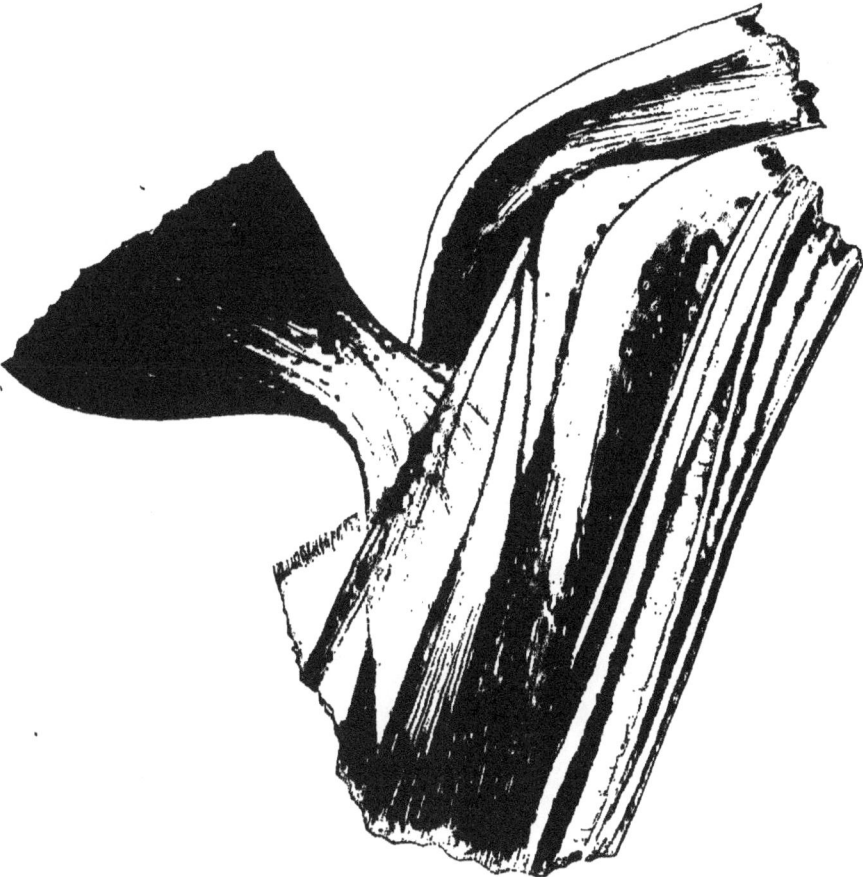

PLATE XXII.

Right upper extremity.

♂, U. S. white, aet. 24.

Gleno-ulnar head, var. 2a, Gleno-ulnar muscle.

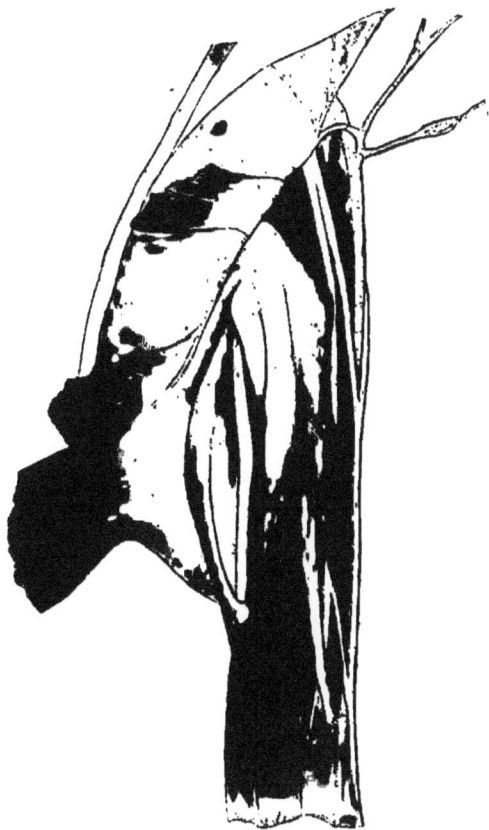

PLATE XXIII.

Right upper extremity.
♂, U. S. white, aet. 50.
Gleno-ulnar head, var. 2a, Gleno-ulnar muscle.

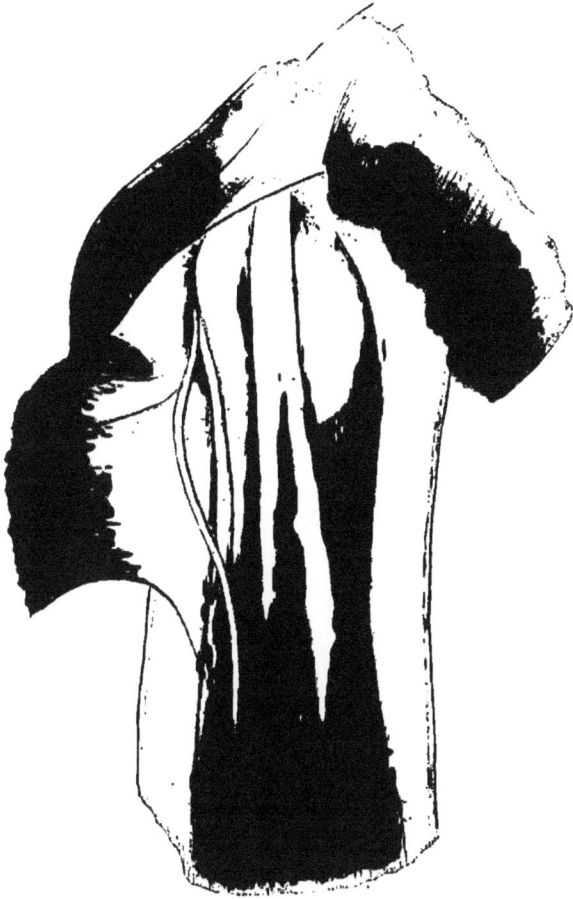

PLATE XXIV.

Right upper extremity.
♂, U. S. negro, aet. 50.
Gleno-ulnar head, var. 2a, Gleno-ulnar muscle.

PLATE XXV.

Left upper extremity.
♀, U. S. white, aet. 23.
Gleno-ulnar head, var. 2a, Gleno-ulnar muscle.

PLATE XXVI.

Left upper extremity.
♂, U. S. negro, aet. 50.
Gleno-ulnar head, var. 2b, Gleno-ulnar muscle.

PLATE XXVII.

Right upper extremity.

♂, Ireland, aet. 32.

Gleno-ulnar head, var. 2b, Gleno-ulnar muscle.

PLATE XXVIII.

Left upper extremity of preceding subject.
♂, Ireland, aet. 32.
Gleno-ulnar, head, var. 2a, Gleno-ulnar muscle.

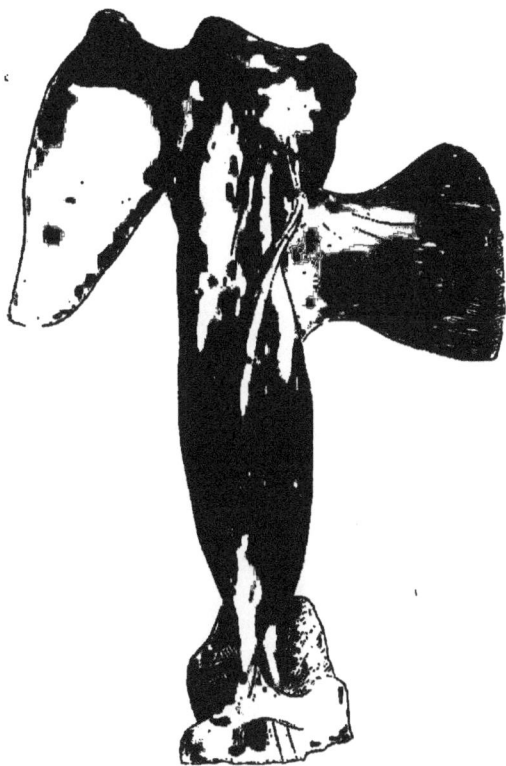

PLATE XXIX.

Left upper extremity.

♂, Ireland, aet. 53.

Gleno-ulnar head, var. 2c, transition forms and variations.

PLATE XXX.

Left upper extremity.
♂, Ireland, aet. 35.
Gleno-ulnar head, var. 2c, transition forms and variations.

PLATE XXXI.

Right upper extremity.
♂, Germany, aet. 64.
Gleno-ulnar head, var. 2c, transition forms and variations.

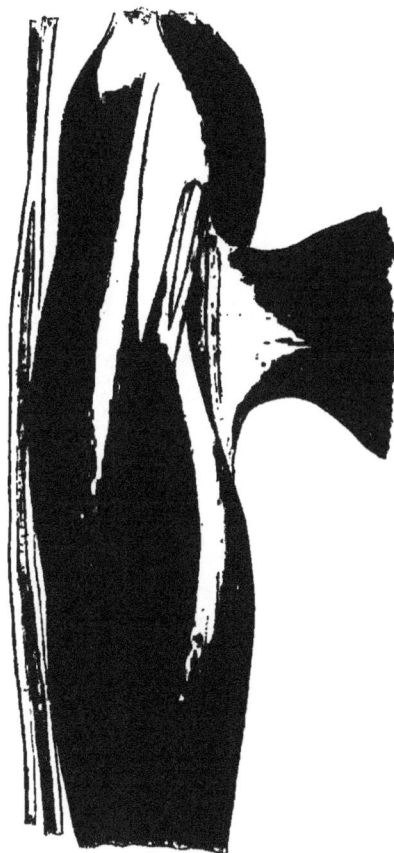

PLATE XXXII.

Left upper extremity.
♂, Germany, aet. 66.
Gleno-ulnar head, var. 3, Gleno-epitrochlear tendon.

PLATE XXXIII.

Left upper extremity.

♂, Ireland, aet. 42.

Gleno-ulnar head, var. 3, Gleno-epitrochlear tendon.

PLATE XXXIV.

Right upper extremity.
♀, Ireland, aet. 27.
Gleno-ulnar head, var. 4, M. Brachio-ulnaris lateralis.

Right upper extremity.
♀, U. S. negro, aet. 24.
Gleno-ulnar head, var. 4, M. Brachio-ulnaris lateralis.

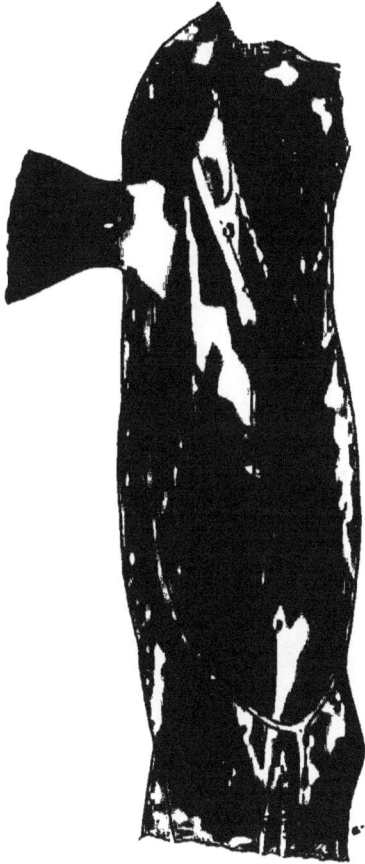

PLATE XXXVI.

Right upper extremity.

♀, Ireland, aet. 72.

Gleno-ulnar head, var. 6, combination of tendinous Gleno-ulnar and Coraco-epitrochlear.

PLATE XXXVII.

Right upper extremity.
♂, U. S. white, aet. 47.
Coraco-ulnar head, var. 2a, Coraco-epitrochlear tendon.

PLATE XXXVIII.

Right upper extremity.

♂, U. S. white, aet. 46.

Coraco-ulnar head, var. 2b, Coraco-epitrochlear and Gleno-epitrochlear tendons combined.

PLATE XXXIX.

Left upper extremity.
♂, Germany, aet. 29.
Coraco-ulnar head, var. 2c; M. Coraco-epitrochlearis.

PLATE XL.

Right upper extremity.
♂, Ireland, aet. 45.
M. Quadriceps flexor cubiti.

PLATE XLI.

Right upper extremity.
♂, Ireland, aet. 40.
M. Quadriceps flexor cubiti.

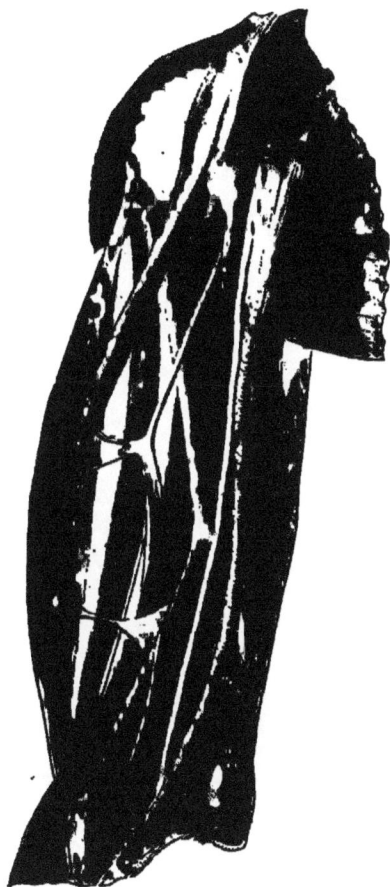

PLATE XLII.

Right upper extremity.
♂, Germany, aet. 62.
M. Quadriceps flexor cubiti.

PLATE XLIII.

Right upper extremity.
♂, U. S. white, aet. 63.
M. Quadriceps flexor cubiti.

PLATE XLIV.

Left upper extremity.
♀, U. S. white, aet. 26.
M. Quadriceps flexor cubiti.

PLATE XLV.

Right upper extremity.

♂, Assyria, aet. 28.

M. Brachialis accessorius, joining insertion of Biceps.

PLATE XLVI.

Left upper extremity.

♂, Ireland, aet. 67.

Gleno-uluar and Coraco-ulnar muscles, complete form, with distal portions persistent.

PLATE XLVII.

Right upper extremity.

♀, Ireland, aet. 54.

Connection of Biceps and Brachialis anticus, with complete derivation of semilunar fascia from the latter muscle.

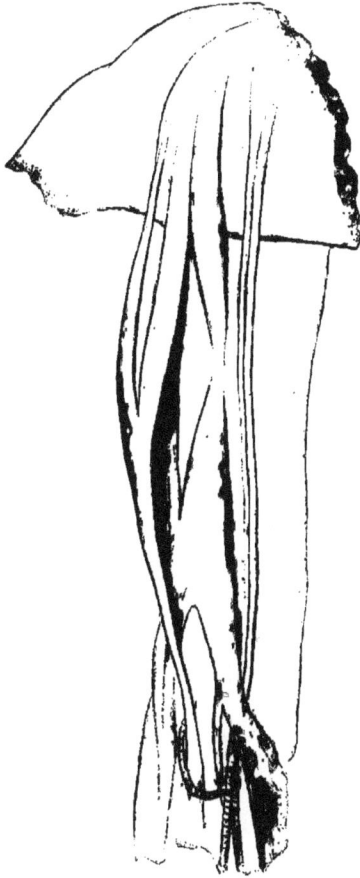